Clinton Hart Merriam

North American Fauna - No. 11

Synopsis of the Weasels of North America

Clinton Hart Merriam

North American Fauna - No. 11
Synopsis of the Weasels of North America

ISBN/EAN: 9783337139599

Printed in Europe, USA, Canada, Australia, Japan

Cover: Foto ©berggeist007 / pixelio.de

More available books at **www.hansebooks.com**

Bridled Weasel, Putorius frenatus.
Valley of Mexico.

Black-footed Ferret, Putorius nigripes.
Western Kansas.

U. S. DEPARTMENT OF AGRICULTURE

DIVISION OF ORNITHOLOGY AND MAMMALOGY

NORTH AMERICAN FAUNA

No. 11

[Actual date of publication June 30, 1896]

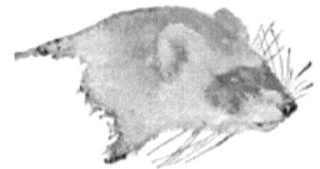

SYNOPSIS OF THE WEASELS OF NORTH AMERICA

C. HART MERRIAM

WASHINGTON

GOVERNMENT PRINTING OFFICE

1896

LETTER OF TRANSMITTAL.

U. S. DEPARTMENT OF AGRICULTURE,
Washington, D. C., May 9, 1896.

SIR: I have the honor to transmit herewith for publication, as No. 11 of North American Fauna, a Synopsis of the Weasels of North America.

Respectfully,   C. HART MERRIAM,
Chief of Division of Ornithology and Mammalogy.

Dr. CHAS. W. DABNEY, Jr.,
Acting Secretary of Agriculture.

CONTENTS.

ILLUSTRATIONS.

(All natural size.)

PLATES.

Frontispiece. Heads of Black-footed **Ferret** and Bridled **Weasel.**

1. Skulls of *Putorius nigripes* and *P. putorius.*
2. **Skulls** of *Putorius arcticus, alascensis, cicognani, streatori,* and *rixosus.*
3. **Skulls** of *Putorius frenatus, longicauda,* and *tropicalis.*
4. **Skulls** of *Putorius noveboracensis, washingtoni,* and *peninsula.*
5. **Skulls** of ***Putorius*** *longicauda, cicognani, noveboracensis, rixosus, peninsula,* and *arcticus.*

TEXT FIGURES.

1. *Putorius nigripes,* ♂ old. Trego County, Kans.
2,3. *Putorius cicognani,* ♂ ad. Elk River, Minnesota.
4-6. *Putorius noveboracensis,* ♂ ad. Adirondacks, New York.
7-9. *Putorius longicauda,* ♂ ad. Fort Sisseton, S. Dakota.
10,11. *Putorius longicauda spadix,* ♀ . Elk River, Minnesota.
12-14. *Putorius arizonensis,* ♂ ad. Boulder County, Colo.
15. *Putorius frenatus,* ♀ ad. **Cofre de** Perote, Vera Cruz, Mexico.
16. *Putorius tropicalis,* ♀ ad. **Jico,** Vera Cruz, Mexico.

4

SYNOPSIS OF THE WEASELS OF NORTH AMERICA.

By C. Hart Merriam.

The present synopsis includes the one ferret and all of the weasels yet discovered in North America north of Panama. Of the true weasels (subgenus *Ictis*) no less than 22 species and subspecies are here recognized, 11 of which are described for the first time.

Until very recently the group has been in a state of chaos, but now, thanks to Outram Bangs's excellent paper entitled 'A review of the weasels of eastern North America,'[1] the obscurity that has so long surrounded our eastern species has been cleared away and the task of revising the whole group is rendered comparatively easy. Additional material is needed from certain parts of the West, particularly from southeastern Alaska and the middle and northern parts of the Great Basin, and much remains to be learned respecting the extent to which intergradation exists between allied forms having contiguous ranges.

Excepting the circumpolar type, represented in America by the weasel of the barren grounds (*Putorius arcticus* nob.), and in Eurasia by the closely related *P. erminea*, the weasels of North America fall naturally into two groups, characterized by important cranial differences, and having complementary geographic ranges. The first is a boreal group comprising five forms: *richardsoni*, *alascensis*, *cicognani*, *streatori*, and *rixosus*, the southernmost of which (*cicognani*) reaches only the northern United States. The other is an austral group comprising the *frenatus* and *longicauda* series and including *P. peninsulæ*, of Florida. Of this series only a single species (*P. arizonensis*) reaches the lowermost of the boreal zones, and this only in the mountains.

Between these two groups are two very interesting species, *noreboracensis* and *tropicalis*—the former inhabiting the eastern United States, the latter the tropical belt of Mexico. Mr. Bangs has already shown that the female of *P. noreboracensis* resembles *P. cicognani*, while the male resembles *P. longicauda*. The case of *P. tropicalis* is exactly parallel, the female resembling *cicognani*, while the male resembles *frenatus*.

[1] Proc. Biol. Soc. Washington, X, pp. 1–24, Feb. 25, 1896.

Among mammals the female is often less specialized than the male
and consequently bears more resemblance to the ancestral stock, thus
giving a clew to the line of descent when this can not be determined
from the male alone. In the present instance the females of *noreboracensis* and *tropicalis* have small, smoothly rounded skulls without sagittal crests and with narrow audital bullæ and inflated squamosals, as
in the *cicognani* series, while the males have large angular skulls with
well-developed sagittal crests, relatively broad audital bullæ, and flat
squamosals, as in the *longicauda-frenatus* series. The inference is that
the austral *longicauda-frenatus* series was derived from the boreal
cicognani stock, and that the differentiation took place in the South.
P. noreboracensis occupies middle ground geographically, and may have
become differentiated from *cicognani* under existing conditions in the
area it now inhabits; but *P. tropicalis*, which inhabits tropical Mexico,
must either have originated from the *cicognani* stock when the latter
was driven southward by the cold of the Glacial epoch, or must have
accomplished a very remarkable migration.

Turning now to the weasel of the tundras (*P. arcticus*), the female is
also found to resemble the *cicognani* type, indicating—at least so far
as the American species go—that the whole group (subgenus *Ictis*) has
sprung from an ancestral type related to *P. cicognani*.

Probably *cicognani* itself is a strongly specialized type, although the
specialization took place a long time ago and seems to have been in
the direction of greater simplicity. The tendency has been toward a
narrowing of the skull as a whole and the obliteration of its promi-
nences and angles. The zygomata have been reduced and drawn in
close to the sides of the cranium, and the brain case has been nar-
rowed, elongated, and smoothly rounded off, as if to enable the head to
pass through small openings. The body as a whole has undergone
parallel modification, presenting the extreme degree of slenderness
known among the mammalia. This type of weasel seems to have been
developed for the express purpose of preying upon field mice or voles,
its narrow skull and cylindrical body enabling it to enter and follow
their runways and subterranean galleries. The extreme development
of the type is presented in *P. rixosus* and *P. streatori*, whose exceedingly small size and almost serpentine form make it possible for them
to traverse the burrows of even the smaller mice.

It is an interesting fact that the geographic range of the *cicognani*
group is almost coincident with that of the field mice of the subgenus
Microtus. Farther south, where these mice occur sparingly or not at
all, the *cicognani* series of weasels is replaced by the larger and more
powerful *longicauda-frenatus* series. Where the ranges of the two
overlap, as on the northern plains, the large weasel (*P. longicauda*)
preys chiefly on pocket gophers (*Thomomys* and *Geomys*) and ground
squirrels (*Spermophilus franklini* and *S. 13-lineatus*), while the smaller
species (*cicognani* and *rixosus*) prey chiefly on mice.

Similarly in the far North, where the frozen tundras are inhabited by lemmings as well as voles, two weasels are present: the tiny, narrow-skulled *rixosus*, which feeds mainly on mice, and the large, broad-skulled *arcticus*, which feeds chiefly on lemmings and rabbits.

It seems clear, therefore, that the different types of weasels have been developed by adaptation to particular kinds of food.

It is much to be regretted that specimens of the South American weasels are not available for study in connection with the North American species. The only one I have seen is *P. affinis* Gray, which ranges from Costa Rica to northern South America. While differing specifically from *frenatus* it clearly belongs to the same group.

Except in winter, weasels are usually so difficult to procure in anything like satisfactory series that but few are available from most of the localities represented in collections. As a rule, the number is too small to afford reliable average measurements; hence the averages here given are subject to correction.

The skull drawings in Pl. I and those in the text (except figs. 10, 11, 15, and 16) were made by Benjamin Mortimer. Those in Pls. II to V, inclusive, were drawn by Dr. James C. McConnell under the supervision of the author. About half of the skulls shown in the latter plates were used by Mr. Bangs in his paper already referred to.

Except where the contrary is distinctly stated, all the measurements in this paper were taken in the flesh by the collector. It is hardly necessary to add that all measurements are in millimeters.

Genus PUTORIUS Cuvier, 1817.

Key to subgenera (for American forms only):
Size large, about equaling the mink (*Lutreola*); facial bar black; legs and feet abruptly darker than upper parts.........................subgenus *Putorius*.
Size medium or small, never more than half as large as the mink (*Lutreola*); facial bar white or absent; legs and feet concolor with or paler than upper parts...subgenus *Ictis*.

Subgenus PUTORIUS Cuvier, 1817.

Putorius Cuvier: Règne Animal, I, 147-149, 1817.
Cynomyonax Coues: Fur-Bearing Animals, 99, 117-148, 1877.

PUTORIUS NIGRIPES Aud. & Bach. Black-footed Ferret.

(Pl. I, figs. 1, 1a, 1b.)

1851. *Putorius nigripes* Aud. & Bach.: Quadrupeds N. Am., Vol. II. pp. 297-299, pl. 93, 1851.
1877. Coues: Fur-Bearing Animals, 149-153, 1877.

Type locality.—Plains of the Platte River, in Nebraska.

Geographic range.—Great Plains, from western North Dakota and northern Montana to Texas; not known west of eastern base of Rocky Mountains.

Characters.—Size of the mink; ears rather large; color buffy, with a dark area in middle of back; fore and hind feet, end of tail, and band across face (including eyes) black.

Color.—Ground color pale yellowish or buffy above and below, clouded on top of head (and sometimes on neck also) by dark-tipped hairs; face crossed by a broad band of sooty black, which includes the eyes; feet, lower part of legs, terminal third of tail, and preputial region, sooty black; back, about midway between fore and hind legs, marked by a large patch of dark umber-brown, which fades insensibly into the buffy of surrounding parts; muzzle, lips, chin, a small spot over each eye, a narrow band behind black facial bar, and sides of head to and including ears, soiled white; anterior margin of ear near base clouded with dusky.

Cranial characters.—Skull large and massive, very broad between orbits, and deeply constricted behind postorbital processes,[1] which are strongly developed; zygomata strongly bowed outward; audital bullæ obliquely flattened on outer side; a prominent bead over lachrymal opening.

Compared with our American weasels, the skull of *Putorius nigripes*

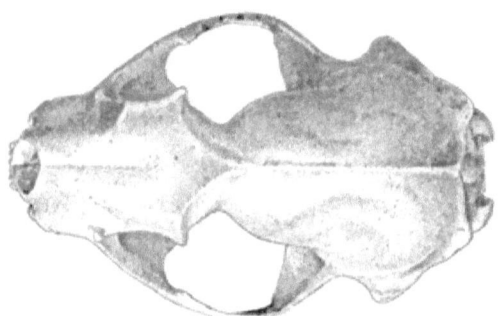

may be told at a glance by its great size, the basilar length in adult males averaging about 65 mm., and in females about 62 mm. Compared with *P. eversmanni* of southern Siberia, it may be distinguished by the greater postmolar production of the palate, and by other minor cranial characters. From

FIG. 1—*Putorius nigripes* ♂ ad. Trego County, Kans.

the common polecat of Europe (*Putorius putorius*) it differs in several important characters, as may be seen by reference to Pl. I. In *P. putorius* the postorbital region is very broad, the postmolar part of the palate exceedingly long, and the anterior part of the audital bullæ very different.

Remarks.—The black-footed ferret bears no resemblance whatever to any other American mammal, but is very closely related to the Siberian *Putorius eversmanni*. It differs from the latter in having much shorter and coarser fur, larger ears, and longer postmolar extension of the palate.

In some specimens of *Putorius nigripes* the pale buffy of the under parts is clouded across the breast between the fore legs, suggesting the dark breast of *P. eversmanni*. The dark facial mask encircles the eyes

---

[1] This constriction deepens with age, as in all the weasels. It is very deep in the skull shown in the accompanying text figure (fig. 1), which is that of an old individual; much less deep in the younger specimen shown on Pl. I, fig. 1.

(including the whitish supraorbital spot) and dips slightly forward before passing transversely across the face, so that its posterior border is in front of the plane of the outer angles of the eyes. Its anterior border sometimes extends forward almost to the nasal pad, but this is unusual. The black of the feet reaches up and covers the fore leg to the elbow, except along the outer side, and the hind leg to near the knee, except posteriorly.

Measurements.[1]—Average of 3 males: Total length, 570; tail vertebræ, 133; hind foot, 60. Average of 2 females: Total length, 500; tail vertebræ, 120; hind foot, 55.

Cranial measurements.—Average of 4 skulls of adult males: Basal length, 64; basilar length of Hensel, 62.5; zygomatic breadth, 43; mastoid breadth, 37; breadth across postorbital processes, 22.5; interorbital breadth, 18; breadth of constriction, 12.5; palatal length, 33; postpalatal length, 31.5. Average of 2 skulls of adult females: Basal length, 60.5; basilar length of Hensel, 58.5; zygomatic breadth, 39; mastoid breadth, 34.5; breadth across postorbital processes, 20; interorbital breadth, 16.5; breadth of constriction, 12; palatal length, 31; postpalatal length, 29.

Subgenus ICTIS Kaup, 1829.

Ictis Kaup: Entwickelungs-Geschichte und Natürliches System der Europäischen Thierwelt, pp. 40–41, 1829. (Contains only a single species, *Mustela vulgaris*.) Schulze: Faunæ Saxonicæ, Mammalia, p. 170, 1893.

Arctogale Kaup: Entwickelungs-Geschichte und Naturliches System der Europäischen Thierwelt, p. 30, 1829. (Contains two species, *erminea* and *boccamela*.)

Gale Wagner: Supplement Schreber's Säugthiere, II, p. 234, 1841. (Contains four species, *frenatus, erminea, boccamela,* and *vulgaris.*)

The names *Ictis* and *Arctogale* were proposed simultaneously in the same publication. Each is accompanied by a diagnosis and included species. The two names, therefore, according to Canon 18 of the A. O. U. Code of Nomenclature, are equally pertinent. In sequence of pagination *Arctogale* comes 10 pages ahead of *Ictis*. *Ictis* contains a single species (*vulgaris = nivalis* Linn.), while *Arctogale* has two (*erminea* and *boccamela*). The reasons for choosing *Ictis* instead of *Arctogale* are: (1) The type of *Ictis* is fixed beforehand, since it contained only a single species, while in *Arctogale* the type must be established arbitrarily; (2) *Arctogale* is now in current use for another genus of small carnivora;[2] to transfer it to a different group would lead to much confusion, and would be a great and seemingly unnecessary calamity. Hence, since there is no rule to the contrary, the better course seems to be to adopt *Ictis* and allow *Arctogale* to fall into synonymy.

[1] The number of specimens of which reliable flesh measurements are available is too small to afford satisfactory averages.

[2] *Arctogale* Peters, 1864, a genus of Viverridæ; Gray, Proc. Zool. Soc. London, 1864, pp. 508, 542–543; Blanford, Fauna British India, Mammalia, p. 114, 1888; Flower and Lydekker, Introduction to Study of Mammals, p. 533, 1891; Lydekker, Royal Nat. Hist., I, p. 461, 1893–94.

Furthermore, *Ictis* has been already revived by Schulze (Faunæ Saxonicæ, Mammalia, 170, 1893), though used by him in a much more comprehensive sense than that originally intended.[1]

List of North American Weasels with type localities.

No.	Name.	Type locality.
1	*Putorius cicognani*	Northeastern North America (north of lat. 41°)
2	*cicognani richardsoni*	Fort Franklin, Great Bear Lake.
3	*richardsoni alascensis*	Juneau, Alaska.
4	*streatori*	Skagit Valley, Washington.
5	*rixosus*	Osler, Saskatchewan.
6	*arcticus*	Point Barrow, Alaska.
7	*arcticus kadiacensis*	Kadiak Island, Alaska.
8	*noveboracensis*	State of New York.
9	*washingtoni*	Trout Lake, Mount Adams, Washington.
10	*peninsulæ*	Tarpon Springs, Florida.
11	*longicauda*	Carlton House, Saskatchewan.
12	*longicauda spadix*	Fort Snelling, Minn.
13	*saturatus*	Siskiyou Mountains, Oregon.
14	*arizonensis*	Flagstaff, Arizona.
15	*alleni*	Black Hills, South Dakota.
16	*xanthogenys*	Southern California.
17	*xanthogenys oregonensis*	Rogue River Valley, Oregon.
18	*frenatus*	Valley of Mexico.
19	*frenatus goldmani*	Pinabete, Chiapas, Mexico.
20	*frenatus leucoparia*	Patzcuaro, Michoacan, Mexico.
21	*tropicalis*	Jico, Vera Cruz, Mexico.
22	*affinis*	Colombia, South America.

PUTORIUS CICOGNANI Bonap. Bonaparte's Weasel.

(Pl. II, figs. 3, 3a, 4, 4a.)

1829. *Mustela (Putorius) vulgaris* Richardson: Fauna Boreali-Americana, Mammalia, pp. 45–46, 1829.
1838. *Mustela cicognanii* Bonaparte: Iconografia Fauna Italica, I, fasc. XXII, p. 4, 1838; Charlesworth's Mag. Nat. Hist., II, p. 37, Jan., 1838.
1839. *Putorius cicognanii* Richardson: Zoology Beechey's Voyage, p. 10*, 1839.
1857. Baird: Mammals North America, pp. 161–163, 1857.
1891. Mearns: Bull. Am. Mus. Nat. Hist., N. Y., III, p. 235, May, 1891.
1896. *Putorius richardsoni cicognani* Bangs: Proc. Biol. Soc. Wash., X, pp. 18–21, Feb. 25, 1896.
1877. *Putorius vulgaris* Coues: Fur-Bearing Animals, pp. 102–109, 1877. Merriam: Mammals Adirondacks, pp. 54–56, 1882 (habits); and most recent authors.

Type locality.—Northeastern North America.

Geographic distribution.—Boreal forest covered parts of North America from New England and Labrador to coast of southeastern Alaska (Juneau, Wrangel, and Loring), and south in the Rocky Mountains to Colorado (Silverton). It occurs in the interior of British Columbia (at Sicamous), but in the Puget Sound region is replaced by a smaller and

---

[1] Schulze included in *Ictis* the two European weasels, *vulgaris* and *erminea*, and also the mink, *lutreola*, and polecat, *putoria*.

darker form. *P. streatori.* In the United States it is common in New England and New York, and in the forest-covered parts of Minnesota. It probably occurs also in northern Michigan and Wisconsin.

General characters.—Size small; tail slender and rather short; color of under parts covering toes and inner sides of both fore and hind feet; color of upper parts never encroaching on belly, but ending along a straight line.

Color.—Upper parts in *summer pelage:* uniform dark brown, hardly darker on head; end of tail blackish; no dark spot behind corners of mouth; under parts, usually including upper lip, white, more or less tinged with yellow. In *winter pelage:* pure white with a strong yellowish tinge on rump, tail, and under parts; end of tail black.

Cranial characters.—Skull small, light, narrow, and elongated without marked postorbital processes, and only a slight postorbital constriction; zygomata narrow, and not bowed outward; brain case elongate and subcylindric; audital bullæ small, narrow, and subcylindric, almost continuous anteriorly (except in old age) with the greatly inflated squamosals; palate narrow; the tooth rows more nearly parallel than in the other species; skull of female similar to that of male, but smaller. Contrasted with *richardsoni,* the skull of *cicognani* is smaller, the audital bullæ decidedly smaller, and the dentition lighter. In nearly every series of *cicognani* there are one or two old males whose skulls are abnormally large and closely resemble skulls of *richardsoni,* except that the audital bullæ are always smaller.

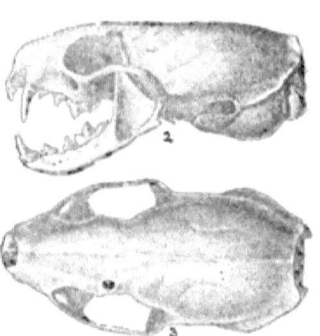

FIGS. 2 and 3.—*P. cicognani* ♂ ad. Elk River, Minnesota.

Measurements.—Average of 5 males from Ossipee, N. H.: Total length, 278; tail vertebræ, 80; hind foot, 36.5. Average of 3 females: Total length, 230; tail vertebræ, 69; hind foot, 30.5.

PUTORIUS CICOGNANI RICHARDSONI (Bonap.). Richardson's Weasel.

1829. *Mustela (Putorius) erminea* Richardson: Fauna Boreali-Americana, pp. 46-47, 1829. (In part: specimen from Fort Franklin, Great Bear Lake. Not *M. erminea* Linn.)

1838. *Mustela richardsoni* Bonap.: Charlesworth's Mag. Nat. Hist., Vol. XI, p. 38, 1838. (based on Richardson's specimen from Great Bear Lake).

1839. *Putorius richardsoni* Rich.: Zool. Beechey's Voyage of Blossom, Mammalia, 10 *, 1839.

1896. Bangs: Proc. Biol. Soc. Washn., X, pp. 1-24, Feb. 25, 1896. (In part.)

Type locality.—Fort Franklin, Great Bear Lake.

Geographic distribution.—Hudsonian timber belt from Hudson Bay to interior of Alaska and British Columbia.

General characters.—Similar to *P. cicognani* but larger; tail of medium length, its terminal third black.

Color.—Upper parts dull chocolate brown, this color reaching down on both fore and hind feet to base of toes; underparts whitish, more or less suffused with yellowish, the pale color extending out in a very narrow and sometimes interrupted strip along inner side of hind feet to toes; tail concolor all around except at tip, which is black for about one-third the total length of tail. In *winter pelage:* white all over except terminal third of tail, which is black; rump and belly more or less tinged with yellowish.

Cranial characters.—Skull long, narrow, and subcylindric like that of *cicognani*, from which it differs chiefly in larger size, larger audital bullæ, and heavier dentition.

Remarks.—*P. richardsoni*, as pointed out by Mr. Bangs, is simply a more northern form of *cicognani*, with which it intergrades completely. It inhabits the Hudsonian timber zone while *cicognani* inhabits the Canadian. On the north, where the timber ends and the tundra begins, the range of *richardsoni* meets that of *arcticus*. The two species differ widely in both cranial and external characters. The light subcylindric skulls of *richardsoni*, with the narrow frontals and appressed zygomata, require no comparison with the broad massive skulls of *arcticus* with their broadly flattened frontals and widely spreading zygomata. The external differences are almost as marked. In *richardsoni* the underparts are nearly white or, at most, only tinged with pale yellowish; the color of the upper parts covers both fore and hind feet, reaching the base of the toes; the tail is relatively long, concolor except at the tip, which is black for about one-third its length. In *arcticus* the underparts are deep yellow; the color of the upper parts stops short of the fore feet and reaches only halfway down the hind feet; the tail is short, yellow below on its basal half, and has a long, black pencil covering at least half its entire length.[1]

Measurements.—(From dry skin of male from Fort Simpson): Total length, 390; tail vertebræ, 95; hind foot, 43 (probably 45).

PUTORIUS RICHARDSONI ALASCENSIS subsp. nov. Juneau Weasel.

(Pl. II, figs. 2, 2a.)

Type from Juneau, Alaska, No. 74423, ♂ ad., U. S. National Museum, Dept. Agric. coll. Collected August 22, 1895, by Clark P. Streator. Original number 4806.

General characters.—Similar in size and general appearance to *P. richardsoni*, but white tips of fore and hind feet more extensive and interorbital region very much broader.

Color.—Upper parts dull chocolate brown, this color reaching down on fore legs to wrists and on hind legs to middle of upper side of feet;

---

[1] It is not strange that Mr. Bangs failed to discriminate between *arcticus* and *richardsoni*. The available material is scanty and mostly of poor quality, and most of the skins had the skulls inside. Through the kindness of Mr. F. W. True, curator of mammals in the United States National Museum, the skulls have been removed and placed at my disposal.

terminal third of tail black; under parts, including upper lip, fore feet, and distal half of hind feet, soiled white, tinged with yellowish. Winter pelage probably white.

Cranial characters.—Skull similar to that of *P. richardsoni*, but very much broader between orbits and across muzzle; postorbital processes more strongly developed; constriction deeper.

Remarks.—Mr. Streator obtained two males of this new weasel at Juneau in the latter part of August. He obtained also, at the same place and time, three females, which in color and markings agree with the males, but are hardly half as large. Their skulls are as small as those of true *cicognani*, which they closely resemble. If they are the females of *alascensis*, as seems probable, then this weasel exhibits as great sexual difference in size as *P. noveboracensis*, in which respect it stands unique as a member of the *cicognani* group. The only alternate possibility is that *cicognani* and *alascensis* occur together at Juneau, and that of the 5 specimens collected there by Streator the 2 males are *alascensis* and the 3 females *cicognani*.

Measurements.—Average of two males from Juneau, Alaska: Total length, 335; tail vertebræ, 95; hind foot, 48. Average of three females from same place: Total length, 270; tail vertebræ, 77; hind foot, 34.

PUTORIUS STREATORI sp. nov.   Puget Sound Weasel.

(Pl. II, figs. 5, 5a, 6, 6a.)

Type from Mount Vernon, Skagit Valley, Washington. No. 76646, ♂ ad., U. S. Nat. Mus., Dept. Agric. coll. Coll. Feb. 29, 1896, by D. R. Luckey. (Original number 3)

Geographic distribution.—Puget Sound and coast region of Washington and Oregon; south at least to Yaquina Bay (Newport), Oregon. Confined to a narrow strip along the coast.

General characters.—Similar to *Putorius cicognani*, but smaller and darker, with color of upper parts encroaching on belly.

Color.—Upper parts, including upper lip and fore and hind feet, uniform dark chocolate brown, darkest on head, and encroaching far on belly and throat (often meeting along middle of belly); terminal third of tail black; under parts narrowly and irregularly white, faintly tinged with yellowish. *In winter pelage* at low altitudes the color of the upper parts is paler (almost drab brown) and the toes may become white; at higher altitudes the whole animal changes to white,[1] except the end of the tail, which always remains black.

Cranial characters.—Skull of male similar to that of male *cicognani*, but smaller, slightly broader interorbitally, and with somewhat more

[1] Mr. R. E. Darrell, of Port Moody, British Columbia, writes me: "I have discovered that, although the weasels do not change color down near salt water, they do change to the white winter coat in the mountains." Specimens in the Department collection from Mount Adams, Washington, killed in February and March, are in the white winter pelage. The type and a female from the same locality (Mount Vernon, Skagit Valley) are in the drab-brown winter pelage.

prominent postorbital processes and smaller audital bullæ. Skull of female very much smaller and more delicate than that of male, resembling female of *cicognani*, but smaller.

Remarks.—*Putorius streatori* is a dark Pacific Coast form of *cicognani*, with which it may be found to intergrade. It differs conspicuously from *cicognani* in the color of the under parts, the dark chocolate brown of the back and sides encroaching far on the throat and usually meeting along the median line of the belly, thus reducing the white to a narrow and irregular strip, which expands on the anterior part of the throat, on the breast behind the fore legs, and immediately in front of the hind legs, and stops abruptly on the under surface of the thighs.

Five winter specimens from Sumas, British Columbia, kindly loaned by Mr. Outram Bangs, point toward intergradation with *cicognani*. In three out of the five, the toes of both fore and hind feet are white, and the color of the upper parts is much paler than in summer pelage. Two of these specimens have the bellies broadly white, as in *cicognani*. They are also much larger than *streatori*. Specimens from Sicamous, in the interior of British Columbia, are fairly typical *cicognani*, having the under parts broadly white; the upper lip, a strip along the inner border of the hind feet, and the toes of both fore and hind feet, white. Specimens from southeastern Alaska (Juneau, Wrangel, and Loring) must also be referred to *cicognani*, and not *streatori*.

Measurements.—Unfortunately, no flesh measurements are available from the type locality. Specimens from Trout Lake, near Mount Adams, Washington, are slightly smaller than the Mount Vernon specimens, and measure as follows: Average of two adult males: Total length, 270; tail vertebræ, 83; hind foot, 33. An adult female: Total length, 210; tail vertebræ, 51; hind foot, 24.

PUTORIUS RIXOSUS Bangs.   Bang's Weasel.

(Pl. II, figs. 7, 7a.)

1857. *Putorius pusillus* Baird: Mammals N. Am., pp. 159–161, 1857. (In part: specimen from Pembina.)
1896. *Putorius rixosus* Bangs: Proc. Biol. Soc. Wash., Vol. X, pp. 21–22, Feb., 1896.

Type locality.—Osler, Saskatchewan, Canada.

Geographic distribution.—Boreal America from Hudson Bay to coast of Alaska (St. Michaels); south to northern Minnesota (Pembina) and Montana (Sun River).

General characters.—Smallest weasel known; tail short and without black tip; only American weasel lacking the black tip.

Color.—*Summer pelage:* Upper parts dark reddish brown; tip of tail not darker; under parts white. In *winter pelage:* Pure white all over, including end of tail.

Cranial characters.—Skull (of type specimen, ♀ ad., No. 642 Bangs' Coll.[1]) very much smaller than the smallest female of any other known

---

[1] I am indebted to Mr. Bangs for the privilege of examining this specimen. Unfortunately, the basioccipital is broken off; hence the basilar length is estimated.

species (total length from occiput to front of premaxillæ, 28.5; basal length, 26.5; zygomatic breadth, 14; length of palate, 11; interorbital breadth, 5.5; breadth across postorbital processes, 7.5; length of audital bullæ, 9.5). The skull is a miniature of *P. cicognani* except that the postorbital processes are more prominent, the brain case more compressed, and there is a distinct sagittal ridge.

Measurements.—Type specimen, female, measured in flesh: Total length, 150; tail vertebræ, 31; hind foot in dry skins, 20–22.

PUTORIUS ARCTICUS sp. nov. Tundra Weasel.

(Pl. II, figs. 1, 1*a*; Pl. V, figs. 6, 6*a*.)

Type from Point Barrow, Alaska. No ¹¹²⁹³ ♂ ad. U. S. Nat. Mus. Collected July 16, 1883, by John Murdoch. Original number, 1672.

Geographic distribution.—Arctic coast and tundras. Specimens examined from Anderson River, Franklin Bay, old Fort Good Hope, lower Mackenzie River, Point Barrow, and St. Michaels.

General characters.—Size large; ears small; tail short but with very long black pencil; underparts yellow (including underside of basal half of tail).

Color—(Type specimen, male adult.) Upper parts, including upper lip, dark yellowish brown; chin white; under parts deep ochraceous yellow, broadly including inner and posterior sides of fore legs, whole of fore feet, distal half and inner side of hind feet, and under side of tail to or nearly to black tip; black tip very long, covering at least half of tail (including long terminal hairs); color of upper parts not encroaching on belly. In *winter pelage*, white all over except long black tip of tail; the white tinged with yellow posteriorly.

Cranial characters.—Skull rather large, broad, and massive; frontal very broad interorbitally; muzzle broad and blunt; postorbital processes moderately developed; postorbital constriction marked; zygomata strongly bowed outward; brain case subtriangular and rather short; audital bullæ subcylindric; postglenoid space smaller than in *richardsoni* and hardly inflated except in female. Contrasted with *P. richardsoni*, the skull of *P. arcticus* is somewhat larger, much broader, and more massive; brain case subtriangular instead of subcylindric; zygomata bowed far outward instead of appressed; postorbital processes more prominent; postorbital constriction much deeper; frontal much broader interorbitally; palate broader posteriorly; dentition heavier. Adult male skulls of *P. arcticus* resemble certain old males of *washingtoni*, but differ in much greater breadth of frontal between orbits, broader muzzle, and blunter postorbital processes. *P. arcticus* resembles true *erminea* of Sweden much more closely than it does any American species.

Remarks.—*Putorius arcticus*, which has been heretofore confounded with *erminea* or *richardsoni*, is one of the most strongly characterized species of the genus. It is a large animal with deep ochraceous yellow under parts and a rather short tail which ends in a remarkably long black pencil. The skull differs from all other American weasels in the great breadth of the frontal region and the breadth and bluntness of the muzzle, in both of which respects it resembles true *erminea*. The only American species whose skull approaches it at all is *P. washingtoni*, as mentioned above. In external characters the differences are too great to require comparison.

It is interesting to find in this country an Arctic circumpolar weasel which, though specifically distinct, is strictly the American representative of the Old World *erminea*. The pattern of coloration, as described above (under *color*), is precisely as in *erminea*, but the tints differ materially. The upper parts in *erminea* lack the golden brown of *arcticus*, and the under parts are very much paler and of a different tint, being pale sulphur yellow instead of ochraceous. Moreover, *arcticus* lacks the whitish border to the ear which is present in *erminea*. In winter pelage the two seem to be indistinguishable except by cranial characters.

A small form of *arcticus* occurs on Kadiak Island, Alaska. It has smaller and narrower audital bullæ, less spreading zygomata, less divergent tooth rows, and decidedly shorter postmolar production of palate. It is probably worthy of recognition as subspecies *kadiacensis*. An adult male (No. 65290) collected April 25, 1894, by B. J. Bretherton, measured in the flesh: Total length, 318; tail vertebræ, 86; hind foot, 44. It is in the white winter pelage, just beginning to change, and the terminal half of the tail is black.

Measurements.—From dry skin of type, male adult, Point Barrow, Alaska: Total length, 380; tail vertebræ, 75; pencil, 55; hind foot, 48 (at least 50 in the flesh).

PUTORIUS NOVEBORACENSIS De Kay. New York Weasel.

(Pl. IV, figs. 1, 1a, 2, 2a; Pl. V, figs. 3, 3a.

1840. *Putorius noveboracensis* De Kay: Catal. Mammalia New York, p. 18, 1840 (*nomen nudum*); Zoology of New York, Mammalia, p. 36, 1842.
1840. Emmons: Rept. Quadrupeds Massachusetts, p. 45, 1840.
1857. Baird: Mammals N. Am., pp. 166–169, 1857.
1896. Bangs: Proc. Biol. Soc. Wash., X, pp. 13–16, Feb. 25, 1896.
1877. *Putorius* (*Gale*) *erminea* Coues: Fur-Bearing Animals, pp. 109–136 (in part), 1877.
Putorius erminea Thompson, Aud. & Bach. (part), Allen, Merriam, and most recent authors.

Type locality.—New York State.

Geographic distribution.—Eastern United States from southern Maine to North Carolina, and west to Illinois.

General characters.—Male large; female small; tail long and bushy, much longer than in *cicognani*, but shorter than in *longicauda;* the black terminal part longer than in any other species except *arcticus*, covering one-third to one-half the tail and measuring 50 to 75 mm. Animal turns white in winter in northern part of range. Extraordinary sexual difference in size and cranial characters.

Color.—Summer pelage: Upper parts, including fore and hind feet and anal region, and often encroaching irregularly on belly, rich dark chocolate brown, sometimes suggesting seal brown; under parts (usually including upper lip) white, more or less washed with yellowish; no yellow on under side of tail or on hind feet, the color of under parts stopping short of ankle. *Winter pelage:* In southern part of range similar to summer pelage, but upper parts paler, nearly drab brown. Northern specimens white all over except terminal third of tail, which is jet black; throat, belly, posterior half of back and tail always suffused with yellowish.

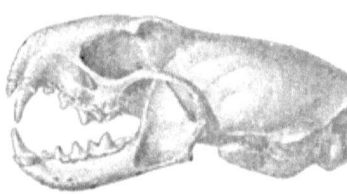

FIG. 4.—*Putorius noveboracensis ♂* ad. Adirondacks, New York.

*Cranial characters.—*Skull of male large, heavy, and elongate; sagittal ridge present in adults; postorbital processes and constriction moderately developed; zygomata *not bowed outward;* audital bullæ rather narrowly oval, usually rounded anteriorly as well as posteriorly. Skull of female very small, light, and narrow, with brain case elongate and subcylindric, much as in *cicognani;* audital bullæ small, narrow, and not rising abruptly anteriorly from inflated squamosals, which latter are elongated and strongly inflated as in *cicognani.* Skulls of males may be distinguished from those of male *longicauda* by shorter postorbital processes, less marked postorbital constriction, less triangular brain case, lower sagittal ridge, very much narrower zygomata, which *are not bowed outward,* narrower palate, and narrower audital bullæ, which are more rounded anteriorly. The resemblance to *P. washingtoni* is very much closer, but male skulls of *novebo-*

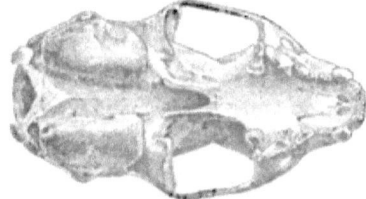 

FIGS. 5 and 6.—*Putorius noveboracensis.* Adirondacks, New York.

racensis may be distinguished by larger size and much larger audital bullæ. The female skull, owing to the inflation of its squamosals inferiorly, needs no comparison with either *washingtoni* or *longicauda,* but is with difficulty separated from *cicognani* in regions where the two species overlap. The postorbital processes are longer and the carnassial and sectorial teeth larger in the females of *noveboracensis* than in *cicognani* from the same localities.

Remarks.—Putorius noveboracensis may usually be distinguished from *P. cicognani* by larger size and also by the longer and more bushy tail, and greater length of the black terminal part. Females of *noreboracensis*, however, sometimes resemble males of *cicognani* rather closely. They may be distinguished not only by the greater length of the tail but also, if in summer pelage, by the absence of yellow from the under side of the tail and inner sides of the hind feet, which parts in *cicognani* usually show more or less yellow.

Measurements.—Average of 10 males: Total length, 407; tail vertebræ, 140; hind foot, 47. Average of 10 females: Total length, 324; tail vertebræ, 108; hind foot, 34.5.

PUTORIUS WASHINGTONI sp. nov. Washington Weasel.

(Pl. IV, figs. 3, 3a, 4, 4a.)

Type from Trout Lake, base of Mount Adams, State of Washington. No. 76522, ♂ ad., U. S. Nat. Mus., Dept. Agriculture collection. Collected December 13, 1895, by D. N. Kaegi.

General characters.—Similar to *P. noreboracensis* in size and general appearance, but with longer tail and shorter black tip. Female very much smaller than male, as in *noreboracensis*.

Color.—Color in summer pelage unknown (probably dark chocolate brown). There are two winter pelages, probably dependent on altitude. In *drab* winter pelage: Upper parts uniform drab brown; end of tail black; under parts white, more or less suffused with pale yellowish. The color of the upper parts encroaches on the sides of the belly as in *noreboracensis*, and a brown spot is present behind the corners of the mouth, which may or may not be confluent with the brown of the cheeks. In the type and two other specimens the hind legs and feet are the same color as the upper parts except that the toes are tipped with whitish and the tips of the fore feet are white. In another specimen, collected January 22, the white is more extensive, covering all of the fore feet and about half of the hind feet. In summer pelage the legs and feet are doubtless the same color as the upper parts, the white of the belly stopping high up on the thighs. In *white* winter pelage: White all over except black tip of tail; tail, rump, and belly strongly suffused with yellow. In one specimen (No. 76604, male, February 7, 1896) the yellow reaches forward over the back nearly to the shoulders; in another (No. 76588, male, February 4, 1896) the whole back is white.

Cranial characters.—The skulls of the two sexes differ greatly: that of the male resembles *noreboracensis* closely in size and general characters, but differs in having the audital bullæ much shorter and the postorbital processes less strongly developed. The postorbital constriction is equally marked. The skull of the female is very much smaller than that of the male, averaging about 38 mm. in length, while the male averages 45 mm. Contrasted with the female of *noreboracensis* the brain case is broader posteriorly and less cylindric. The audital bullæ are more sharply separated from the squamosal inflation and the latter is only slightly marked, not reaching the plane of the bullæ. The resemblance therefore to *P. cicognani* is much less marked in the female *washingtoni* than in the female *noveboracensis*.

Remarks.—This new species is represented in the collection by 14 skulls and 6 skins, of which the greater number are males. The female is darker than the males, and the top of the head is darker anteriorly than the rest of the upper parts, while in the males it is concolor with the back. These differences are probably seasonal, the female not having completed the change from summer to winter pelage, though collected December 11. All are from the Mount Adams region.

Measurements.—The skins, which are well made, afford the following approximate measurements: Male, total length, 240; tail vertebræ, 155; hind foot, 44. Female, total length, 360; tail vertebræ, 120; hind foot, 37.

PUTORIUS PENINSULÆ Rhoads. Florida Weasel.
(Pl. IV, figs. 5, 5a; Pl. V, fig. 5.)

Putorius peninsulæ Rhoads: Proc. Acad. Nat. Sci. Phila., June 1894, 152–155.
Bangs: Proc. Biol. Soc. Wash., X, pp. 10–13, Feb. 25, 1896.

Type locality.—'Hudsons,' 14 miles north of Tarpon Springs, Fla.

Geographic distribution.—Peninsula of Florida; limits of range unknown.

General characters.—Size rather large, about equaling male of *Putorius noveboracensis;* skull similar to that of *longicauda,* but with very large audital bullæ.

Color.—Upper parts dull chocolate brown, darkest on head; upper lip and chin whitish; rest of under parts, including fore feet and toes of hind feet, yellowish; a brown spot behind corners of mouth; a small tuft of white hairs under anterior root of ear. The color of the under parts covers the belly broadly and is not encroached upon by the color of the upper parts. Irregular and inconstant white markings are sometimes present between and behind the eyes.

Cranial characters.—Skull rather massive, resembling that of *longicauda,* but with higher sagittal crest; less spreading zygomata; narrower, higher, and more swollen audital bullæ, and less prominent postorbital processes. Contrasted with *P. noveboracensis* the postorbital constriction is deeper, the brain case higher and more subtriangular, the audital bullæ higher and more swollen, the upper carnassial tooth decidedly larger, and the molar smaller. The upper molar is peculiar: It is short, hardly expanded at either end, and implanted at right angles to the premolar series.

Measurements.—An adult female from Tarpon Springs, Fla.: Total length, 374; tail vertebræ, 127; hind foot, 44.5.

PUTORIUS LONGICAUDA Bonaparte. Long-tailed Weasel.
(Pl. III, figs. 3, 3a, 4, 4a; Pl. V, figs. 1, 1a.)

1829. *Mustela (Putorius) erminea* Richardson: Fauna Boreali-Americana, pp. 46–47, 1829 (in part: Specimen from Carlton House).
1838. *Mustela longicauda* Bonaparte: Charlesworth's Magazine Nat. Hist. N. S., II, p. 37–38, 1838 (based on Richardson's long-tailed variety of *erminea* from Carlton House).

1839. *Putorius longicauda* Rich.: Zool. Beechey's Voyage of Blossom, p. 10, 1839.
1857. Baird: Mammals N. Am., pp. 169–171, 1857.
1877. Coues: Fur-Bearing Animals, pp. 136–142, 1877.
1896. Bangs: Proc. Biol. Soc. Wash., X, pp. 7–8, Feb. 25, 1896.

Type locality.—Carlton Hoase, on North Saskatchewan River, Canada.

Geographic distribution.—Great Plains from Kansas northward.

General characters.—Size large (adult males averaging about 450 mm.

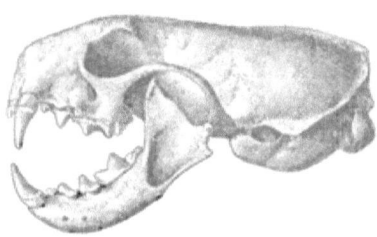

in total length); tail very long (vertebræ 155 mm. or more in males), its black tip rather short; under parts always strongly yellowish or ochraceous.

Color.—Upper parts pale yellowish brown, or pale raw-umber brown, becoming darker on head; terminal part of tail black; chin and upper lip all the way round white; rest of under parts varying

FIG. 7.—*Putorius longicauda.* Fort Sisseton, S. Dak.

from strong buffy yellow to ochraceous orange, the color extending from throat posteriorly, including upper side of fore feet, inner side of hind feet, and upper side of hind toes; under side of tail more or less suffused with yellowish; soles of hind feet brownish. In worn summer pelage the color of upper parts is decidedly paler, and in some old specimens the upper and lower surfaces are not sharply differentiated. The orange tinge of the under parts is strongest on the throat.

Cranial characters.—Skull large, broad, and massive, with well-developed postorbital processes, strongly marked postorbital constriction, and a moderate sagittal crest; zygomata bowed strongly outward; brain case subtriangular as seen from above; audital bullæ rather broad and subrectangular; palate broad;

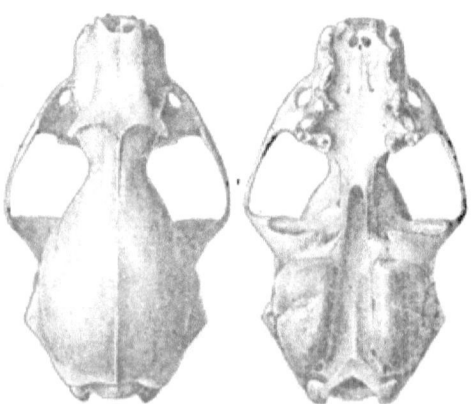

FIGS. 8 and 9.—*P. longicauda* ♂ ad. Fort Sisseton, S. Dak.

dentition heavy; audital bullæ anteriorly rising abruptly from squamosal, which is not inflated in either sex; skull of female similar to male, but smaller, and with only a slight sagittal ridge. Contrasted with male skulls of *noveboracensis* and *washingtoni*, the male of *longicauda* is broader and relatively shorter, with more spreading zygomatic arches, longer postorbital processes, deeper postorbital constriction, and much broader and more rectangular audital bullæ, which as a rule are broadly truncate instead of narrowly rounded anteriorly.

Measurements.—Average of 4 males from plains of Saskatchewan and Alberta: Total length, 450; tail vertebræ, 165; hind foot, 51. Average of 3 females: Total length, 387; tail vertebræ, 144; hind foot, 44.

PUTORIUS LONGICAUDA SPADIX Bangs.

Putorius longicauda spadix Bangs: Proc. Biol. Soc. Wash., X, pp. 8-9, Feb. 25, 1896.

Type locality.—Fort Snelling, near Minneapolis, Minn.

Geographic distribution.—Edge of timber belt in Minnesota, along boundary between Transition and Boreal zones.

General characters.—Similar to *P. longicauda*, but much darker.

Color.—Summer pelage: Upper parts chocolate brown, darkest on the head, but paler than in *noreboracensis:* chin and upper lip whitish all round: rest of under parts, including upper surfaces of fore feet and toes of hind feet, buffy yellow; terminal part of tail black. *Winter pelage:* Snow-white everywhere except black tip of tail and a yellowish suffusion on rest of tail, and sometimes also on under side of hind feet.

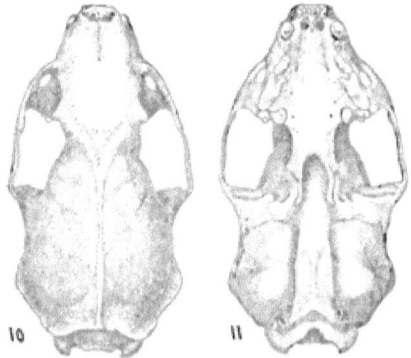

Figs. 10 and 11.—*Putorius l. spadix* ♀ ad. Elk River, Minnesota.

Cranial characters.—As in *P. longicauda.*

Measurements.[1]—Average of 6 males from Fort Snelling, Minn.: Total length, 460; tail vertebra, 166.5; hind foot, 54.5. Average of 3 females: Total length, 356; tail vertebræ, 132; hind foot, 43.5.

PUTORIUS SATURATUS sp. nov. Cascade Mountain Weasel.

Type from Siskiyou, near southern boundary of Oregon (altitude, about 4,000 feet). No. 65930, ♂ ad., U. S. Nat. Mus., Department of Agriculture collection. Collected June 6, 1894, by Clark P. Streator. Orig. No. 3805.

General characters.—Similar to *P. arizonensis*, but larger and darker, with belly more ochraceous, and with distinct spots behind the corners of the mouth.

Color.—Color of upper parts in summer pelage (June) dark raw umber brown, becoming much darker on the top of the head and nose; terminal part of tail black; a brown spot at corner of mouth which may be confluent with brown of cheeks; color of upper parts extending over outer side of forearm to wrist, and over hind foot to toes; chin

---

[1]These measurements were taken in the flesh by Dr. E. A. Mearns, to whom I am indebted for them.

white; rest of under parts ochraceous or orange yellow, including the fore feet, and reaching narrowly down the under side of hind leg to ankle, whence it may or may not extend in a narrow line along inner side of foot to toes; under side of tail more or less suffused with golden chestnut; anal region chestnut brown; in worn pelage the colors are everywhere much paler.

Cranial characters.—Skull similar to that of *P. arizonensis* but with postorbital processes broader at base and less peg like.

Remarks.—This handsome weasel replaces *longicauda* on the Cascade and Siskiyou mountains of Oregon and Washington, reaching a short distance into British Columbia. The only specimens examined have come from Siskiyou, Oregon, and Chilliwack, British Columbia (the latter, No. 3553, collection of E. A. and O. Bangs).

Measurements.—Average of 2 males from Siskiyou Mountains, Oregon: Total length, 423; tail vertebræ, 164; hind foot, 48.

PUTORIUS ARIZONENSIS Mearns. Mountain Weasel.

Putorius arizonensis Mearns: Bull. American Museum Nat. Hist., Vol. III, No. 2, pp. 234–235, May, 1891.
Putorius longicauda Merriam: Mammals of Idaho, N. Am. Fauna, No. 5, pp. 83–84, Aug. 1891 (from mountains of Idaho).

Type locality.—San Francisco forest, Arizona (a few miles south of Flagstaff).

Geographic distribution.—Broadly, the Sierra Nevada and Rocky Mountain systems, reaching British Columbia in the Rocky Mountain region, but not known north of the Siskiyou Mountains in the Sierra-Cascade system.

Fig. 12.—*P. arizonensis* ♂ ad. Boulder County, Colo.

General characters.—Similar to *Putorius longicauda* in color and markings, but much smaller in size.

Color.—Upper parts from occiput to black tip of tail, raw umber brown; head decidedly darker; end of tail black; chin and upper lip all round white; rest of under parts including upper surfaces of fore feet and inner half of hind feet and upper surfaces of hind toes ochraceous or ochraceous yellow, varying in tint.

Cranial characters.—Skull similar to that of *longicauda* but decidedly smaller and less triangular; narrower across mastoids and more bulging in parietals.

Remarks.—*Putorius arizonensis* is a mountain form of *longicauda*, which it closely resembles except in size. The type specimen, collected by Dr. Mearns on the pine plateau of Arizona a few miles south of Flagstaff, is an immature female and is of unusually small size. A male obtained by him near the same place is of the normal size, as is another male in the Department collection from Springerville, Ariz., collected by E. W. Nelson. Specimens from the northern Rocky Mountain region (St. Mary Lake, Montana, and Salmon River and Pahsimeroi Mountains, Idaho) differ in color from the typical animal from Arizona and Colorado, and agree with *alleni* from the Black Hills in having the upper parts strongly suffused with golden brown, the yellow of the under parts yellow rather than ochraceous, and the under side of the tail strongly yellow on the basal half or two thirds. The skulls, however, lack the flattened audital bullæ of *alleni*. Specimens from the Sierra Nevada in California are hardly distinguishable from the Rocky Mountain animal. The only apparent external differences are that the yellow of the under parts reaches up farther under the chin, the white of the upper lip is less extensive, and the under side of the tail is more suffused with yellowish. But none of these characters is constant. In one specimen from Donner, Calif. (No. 2650, female, Merriam Coll.), even the white upper lip is as marked as in Rocky Mountain

specimens; it reaches all the way round, fills the space under the nasal pad to the nostrils, and broadens strongly under the eyes. In cranial characters also the differences are slight and inconstant. The postorbital processes are longer and more slender, often becoming peg-like in old males. The audital bullæ average smaller and more convex anteriorly, and in the female are decidedly narrower and more subcylindric. But in an adult female

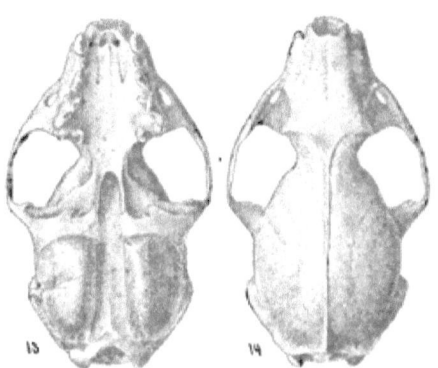

FIGS. 13 and 14.— *P. arizonensis* ♂ ad. Boulder County, Colo.

from Fort Klamath, Oreg., the bullæ are nearly as broad as in Rocky Mountain females. The three female skulls I have seen of the Sierra form are decidedly smaller than females from the Rocky Mountains.

The Sierra specimens show a strong tendency to grade into, or at least toward *xanthogenys*. In nearly half the specimens examined white hairs are present between the eyes, and in several they are sufficiently numerous to form a conspicuous white spot, though the spot is not large and rectangular as in true *xanthogenys*. The white cheek spots I have not seen in Sierra specimens, but the brown spots behind the corners of the mouth are sometimes present (as in No. 30655, male, from Upper Cottonwood Meadows, near Mount Whitney, Calif.).

A specimen from St. George, Utah, an old female, differs in some respects from typical *arizonensis*. The skull is small and relatively short, and the shortening is mainly in the palate and rostral part, which measures 2 mm. less than the average of adult females of *arizonensis* of the same size. Moreover, the postorbital processes are longer and more slender than in any female of *arizonensis* I have examined from either the Rocky Mountain or Sierra systems. Externally the St. George specimen differs from typical *arizonensis* in the following particulars: Yellow of underparts more strongly tinged with ochraceous; white of upper lip narrow and not reaching around anteriorly; brown of upper parts reaching down on outer side of arm to wrist; a small brown spot bearing two bristles just behind each corner of mouth. In this respect, and this only, it resembles *xanthogenys*; there is no trace of white on the cheeks or between the eyes.

Measurements.—Average of 5 males from the Rocky Mountains: Total length, 385; tail vertebræ, 144; hind foot, 44.5. Average of 4 females: Total length, 358; tail vertebræ, 130; hind foot, 40.

PUTORIUS ALLENI sp. nov. Black Hills Weasel.

Type from Custer, Black Hills, South Dakota. No. ⁴¹⁹⁸⁄₅, ♂ ad., Merriam collection. Collected July 12, 1888, by Vernon Bailey. Original No. 90.

Geographic distribution.—Black Hills, South Dakota.

Characters.—Similar to *P. arizonensis* in size and general characters, but upper parts more suffused with yellowish and audital bullæ flatter.

Color.—Upper parts from occiput to black tip of tail golden or yellowish-brown, in some lights with an olivaceous tinge; head dark brown, without yellowish tinge; upper lip and chin white; rest of underparts, including inner sides of legs, whole of fore feet, toes of hind feet and under side of basal part of tail, intense buffy yellow.

Cranial characters.—Skull similar to that of *arizonensis*, but audital bullæ much flatter and somewhat smaller; brain case slightly flatter and bulging laterally immediately behind constriction; frontal somewhat broader interorbitally; skull as a whole shorter. The skull of an old female (No. 7441, Am. Mus. Nat. Hist.) is much smaller than the male, and the audital bullæ are narrow and not flattened. In both sexes the postorbital processes are strongly developed.

Remarks.—*Putorius alleni* is an isolated and only slightly differentiated form of *P. arizonensis*, from which it is completely cut off geographically. It is surrounded on all sides by the large weasel of the plains, *P. longicauda*. In worn summer pelage the color differences that distinguish it from *arizonensis* are not apparent.

I take pleasure in naming the species in honor of Dr. J. A. Allen, of the American Museum of Natural History, New York, who has recently published an important paper on the mammals of the Black Hills, and to whom I am indebted for the loan of three additional specimens.

Measurements (of type specimen, male adult).—Total length, 372; tail vertebræ, 137; hind foot, 44.

PUTORIUS XANTHOGENYS (Gray).   California Weasel.

1843. *Mustela xanthogenys* Gray : Annals and Magazine Nat. Hist., XI, pp. 118, 1843.
1857. *Putorius xanthogenys* Baird : Mammals N. Am., pp. 176-177, 1857.
1877. *Putorius (Gale) brasiliensis frenatus* Coues : Fur-Bearing Animals, pp. 142-146,
      1877 (in part).

Type locality.—Southern California, probably vicinity of San Diego.

Geographic distribution.—Sonoran and Transition faunas of California, on both sides of the Sierra Nevada.

General characters.—Size medium; tail long; face conspicuously marked with whitish, but rest of head not black; under parts ochraceous.

Color.—Upper parts from back of head to terminal part of tail in *summer pelage* raw-umber brown, tinged with golden; in *winter pelage*, drab brown, without yellowish suffusion; head always darker, becoming dusky over nose; a large rectangular spot between eyes, and a broad oblique band between eye and ear, whitish; end of tail black; a brown spot behind corners of mouth; chin white; rest of under parts, including fore feet all round and inner side and toes of hind feet, varying from buffy ochraceous to ochraceous orange. In some specimens the ochraceous covers the greater part of the hind feet as well as the toes.

Cranial characters.—Skull of the *longicauda* type and practically indistinguishable in size and characters from *P. arizonensis*; skull as a whole short and broad; zygomata bowed outward; postorbital processes strongly developed; sagittal ridge distinct; audital bullæ moderate, usually truncate anteriorly; skull of female similar to that of male, but smaller.

Remarks.—*Putorius xanthogenys* inhabits the San Joaquin and Owens valleys and the whole of southern California except the higher mountains. In ascending the mountains it gradually loses the facial markings and seems to grade into *P. arizonensis*, the weasel of the mountain summits.

Measurements.—Average of 7 males from southern California: Total length, 402; tail vertebræ, 156; hind foot, 43.5. Average of 3 females: Total length, 368; tail vertebræ, 135; hind foot, 40.5.

PUTORIUS XANTHOGENYS OREGONENSIS subsp. nov.   Oregon Weasel.

Type from Grants Pass, Rogue River Valley, Oregon. No. $\frac{33242}{11}$, ♀ ad., U. S. Nat. Mus.,
   Dept. Agric. Coll.   Collected December 19, 1891, by Clark P. Streator.   Original
   number 1404.

Geographic distribution.—Rogue River Valley, Oregon; limits of range unknown.

General characters.—Similar to *P. xanthogenys* but decidedly larger, darker in color, and with face markings much restricted.

Color.—Upper parts in winter pelage pale chocolate brown, slightly darker on head; a small and ill-defined patch between eyes, and a nar-

row **vertical** bar between eye and ear, white; throat white; rest of
under parts, including fore feet and inner sides and distal half of hind
feet, pale yellowish; terminal one-fifth of tail black; rest of tail above
and below concolor with back and without the yellowish tinge which
is characteristic of *xanthogenys*.

Cranial characters.—Skull similar to that of *xanthogenys* but larger
and decidedly broader. The skull of the type, an adult female, compared with skulls of *xanthogenys* of the same sex and age from southern California, differs in the following particulars: Skull everywhere
broader; muzzle, palate, interorbital breadth and constriction very
much broader; zygomata more spreading.

Measurements.—Type specimen, female adult: Total length, 412; tail
vertebræ, 155; hind **foot, 44.**

PUTORIUS FRENATUS (Lichtenstein). Bridled Weasel.

(Pl. III, figs. 1, 1a, 1b, 2.)

1813. *Mustela brasiliensis* Sevastianoff: **Mem.** Acad. Imp. **Sci.** St. Petersburg, IV,
	356-363, Table IV, 1813. (Name **on plate** only; diagnosis in text.) Preoc-
	cupied by *Mustela brasiliensis* [an otter] Gmelin, **1788.**
1832. *Mustela frenata* Lichtenstein: Darstellung neuer oder wenig bekannter Sau-
	gethiere, Pl. XLII and corresponding text (unpaged), 1832.
1857. *Putorius frenatus* Baird: Mammals N. Am., 173–176, 1857.

Type locality.—Valley of Mexico, near City of Mexico.

General characters.—Size large; **tail** long; its black tip **relatively**
short; head black, with conspicuous white markings.

Color.—**Top** of head blackish, interrupted between eye and ear by a
broad, whitish band, **which** is nearly confluent with a patch of same
color between the eyes; rest of upper parts brown; a dark spot behind
corners of mouth; **chin and** throat whitish; rest of under parts ochra-
ceous yellow; forefeet **to** or **above** wrists whitish or pale buffy yellowish, continuous with and shading **into** ochraceous of under parts; color
of under parts extending down on **inner** side of hind legs and feet to
toes, which are whitish or yellowish white.

Cranial characters.—Skull large and massive, with strongly **devel-**
oped postorbital processes, deep postorbital constriction, marked sagittal crest, and peculiar audital bullæ, which are obliquely truncated
anteriorly (the inner side reaching farthest forward) and abruptly
highest on inner side, falling away suddenly on outer side so as to
form a rounded ridge along the inner side of the longitudinal axis of
the bulla. The skull of *frenatus* resembles that of *longicauda*, but is
considerably larger, and differs in the form of the audital bullæ just
described, and also in the extent of the postglenoid space, which is
much larger than in *longicauda*. The dentition is heavy and the
upper carnassial tooth relatively shorter than in *longicauda*. The
ramus of the under jaw is much more convex inferiorly.

Remarks.—Lichtenstein, in his original description of *Mustela frenata*,
states that the tail is about one-third longer than **that of the** European

weasel (*erminea*); that only its extreme tip is black; that the head, ears, and crown are black, this coloring fading into the reddish brown of the upper parts on the back of the head behind the ears; that the facial markings, throat, and breast are white; the remainder of the under parts ocher yellow. The white spot between the eyes is described as heartshaped, and in the colored plate it is shown to be nearly, but not quite, confluent with the white patch between the eye and ear. The colors in the plate are not good, as the whole under parts are white instead of ocher yellow, and the black tip of the tail is not shown. The specimen seems to have been in worn pelage. Lichtenstein had two specimens, both collected by Deppe near the City of Mexico.

Fortunately, the Department collection contains two specimens collected by E. W. Nelson at Tlalpam, in the Valley of Mexico, which may be considered topotypes of *frenatus*, for they not only came from the same locality as Lichtenstein's types, but also agree essentially in every detail with his excellent description. The only points in which the description fails to agree absolutely with the specimens is that in the latter the white of the throat is less pure and the black tip of the tail perhaps a trifle more extensive than one would infer from the description; but the throat is white in contrast with the strongly ochraceous yellow of the rest of the under parts, and a specimen in the United States National Museum from the City of Mexico (No. 1060, ♀ ad., J. Potts) has both throat and breast white, as in the original description.

The statement that only the extreme tip of the tail is black was made in comparison with the European weasel (*erminea*), in which nearly half of the tail is black. Hence the description agrees entirely with the specimens in hand. One point not mentioned in the description is shown in the plate, namely, that the hind feet and toes are in large part whitish or yellowish white. The quantity of white is variable. In a young male from Tlalpam (No. 50827) it is restricted to the inner side of the foot, hardly reaching the toes, while in an adult male from the same locality (No. 50826) it includes the toes. The whitish spot between the eyes is also variable, both in form and extent. Lichtenstein described it as heart-shaped, and his figure shows that it is narrow where it approaches closest to the stripe between the eye and ear, with which it is nearly, but not quite, confluent. This is precisely its condition in the adult male from Tlalpam, which may be considered a duplicate type of the species. In this specimen the median white spot is almost divided by the dark color of the forehead, which pushes down between the eyes, so that the whitish spot might be described as a narrow stripe over each eye, the two becoming confluent below. In the young specimen the white spot is subrectangular and not divided by the black of the forehead.

Note on Putorius brasiliensis.—In 1813 a Russian naturalist, Sevastianoff, gave the name '*Mustela brasiliensis*' to a weasel brought to St. Petersburg by Capt. A. J. Krusenstern on his return from a voyage around the world. The animal was said to have come from Brazil, but no definite locality was given. In the numerous publications that have since appeared relating to the mammals of Brazil and adjacent territory, no weasels are mentioned as inhabiting that country, and the species described from the mountains to the westward differ so widely from Sevastianoff's *brasiliensis* that it is almost certain his animal did not come from Brazil. The original description (including measurements) agrees in every respect with *P. frenatus* of Lichtenstien from the Valley of Mexico, indicating that the two animals are identical. On this assumption the well-known and appropriate name *frenatus* would have to fall before the earlier and inappropriate *brasiliensis.* Fortunately, however, Sevastianoff placed his animal in the genus *Mustela*, and the name *Mustela brasiliensis* is preoccupied by Gmelin for a South American otter. (Syst. Nat., ed. 13, p. 93, 1788.) Hence, unless some earlier name is found, *frenatus* will stand for the Mexican bridled weasel.

Measurements.—An adult male from Tlalpam, Valley of Mexico (type locality): Total length, 505; tail vertebræ, 203; hind foot, 53. Average of 6 males from Brownsville, Tex.: Total length, 488; tail vertebræ, 192; hind foot, 51. Average of 3 females from Brownsville: Total length, 438; tail vertebræ, 187; hind foot, 41.5.

PUTORIUS FRENATUS GOLDMANI subsp. nov.

Type from Pinabete, Chiapas, Mexico. No. 77519, ♂ ad., U. S. Nat. Mus., Dept. Agric. coll. Collected Feb. 10, 1896, by E. A. Goldman. Altitude about 8,200 feet (= 2,500 meters). Original number 9279.

Geographic distribution.—Mountains of southeastern Chiapas; limits of range unknown.

General characters.—Similar to *P. frenatus* in size and general characters, but tail and hind feet longer; light markings more restricted; black of head reaching much farther back on neck; color of upper parts darker and more extensive, encroaching on sides of belly and covering fore and hind feet; black tip of tail longer.

Color.—Upper parts, including whole of fore and hind feet, dull, dark chestnut brown, washed with black on the neck from shoulders forward, and becoming pure black on the head; face marked by a whitish patch between the eyes, and a narrow, oblique band between eye and ear; a blackish spot behind angle of mouth; color of under parts salmon ochraceous, reaching wrists inferiorly, but not reaching heels; terminal third of tail black.

Cranial characters.—Skull rather large; zygomata moderately spreading; squamosal inflation moderate, but large for a member of the *frenatus* series; audital bullæ small, steep on inner side, and only slightly elevated anteriorly above squamosal inflation. The skull as a whole resembles that of *frenatus*, but differs conspicuously in the greater length and inflation of the postglenoid part of the squamosal, greater breadth of the basioccipital, and in the size and form of the audital bullæ. The latter are very narrow, low anteriorly where they meet the inflated squamosal without an abrupt step, and high along the inner side.

Remarks.—Mr. E. W. Nelson writes me that this fine weasel is found sparingly in the forest about Pinabete, Chiapas, at an altitude of 7,000 to 8,000 feet (2,100 to 2,500 meters). The type specimen was shot in the afternoon while hunting on a heavily wooded hill slope. It was heard making long, slow leaps over the dry, crisp leaves. Coming to a log, it stood up and rested its fore feet on the log, in which position it was shot by Mr. Goldman.

A specimen from Cerro San Felipe, Oaxaca, is intermediate, both in coloration and cranial characters, between typical *frenatus* and *goldmani*; hence there is little room for doubt that complete intergradation exists between the two.

Measurements.—Type specimen, male adult: Total length, 504; tail vertebræ, 201; hind foot, 58.

PUTORIUS FRENATUS LEUCOPARIA subsp. nov.

Type from Patzcuaro, Michoacan, Mexico. No. ⁴⁵⁸¹⁴, ♂ ad., U. S. Nat. Mus., Dept. Agric. coll. Collected July 27, 1892, by E. W. Nelson. Original number 2860.

General characters.—Similar to *Putorius frenatus*, but slightly larger; black of head extending posteriorly over neck; white face markings much more extensive; the spot between the eyes very much larger and broadly confluent on both sides with whitish area between eye and ear, which area also is much more extensive in all directions than in *frenatus.*

Color.—Upper parts from shoulders to black tip of tail, dark brown; neck, crown of head, nose, ears, and sides of face to a little behind the eye, black; black of head between eyes and ears divided by a broad band of buffy white which is broadly confluent with buffy yellow of throat and chin; a narrow border of whitish on upper lip; rest of under parts ochraceous yellow (including whole of fore feet, inner sides of hind legs and feet, and terminal half or nearly half of upper surfaces of hind feet, where the color becomes paler, being buffy ochraceous, as on the throat).

Cranial characters.—Skull similar to that of *frenatus*, but larger; audital bullæ much narrower; postorbital processes less strongly developed.

Remarks.—This handsome weasel presents the maximum of black and white markings known in the *frenatus* group, the black of the head reaching back over the neck and the white face markings covering a large area. In the type specimen a white stripe 50 mm. in length extends down the middle of the nape from a point between the ears more than halfway to the shoulders. This, however, is probably abnormal, though a trace of it exists in a female from the same locality. This form is the poorest subspecies described in the present paper.

Measurements.—Average of 2 males from Patzcuaro (type locality): Total length, 510; tail vertebrae, 201; hind foot, 53. An adult female from same place: Total length, 400; tail vertebrae, 159; hind foot, 42.

PUTORIUS TROPICALIS sp. nov.   Tropical Bridled Weasel.

(Pl. III, figs., 5, 5a, 6, 6a.)

Type from Jico, Vera Cruz, Mexico   No. 51994, ♂ ad., U. S. Nat. Mus., Dept. Agric. coll.   Collected July 9, 1893, by E. W. Nelson.   Altitude 6,000 feet (1,800 meters).   Original number 5195.

Geographic distribution.—The tropical coast belt of southern Mexico and Guatemala from Vera Cruz southward.

General characters.—Similar to *Putorius frenatus*, but much smaller and darker, with the white face markings less extensive, the belly pale orange instead of ochraceous, and under side of tail very much darker.

Color.—Upper parts deep umber brown with a fulvous tone; head, ears, and neck, black, passing gradually into brown of back just in front of the shoulders; terminal one-fourth (or a little more) of tail, black; face markings as in *frenatus*, but less extensive and whiter; under parts ochraceous buff on throat and fore feet, becoming rich orange buff on belly and inner side of thighs, whence (becoming paler) the color reaches out in a narrow interrupted stripe along the inner side of the hind feet to the toes, which are irregularly buffy.

Cranial characters.—Skull of male similar in general to that of *frenatus*, but smaller, relatively longer, with less spreading zygomata, less strongly developed postorbital processes, and probably broader postorbital constriction (the type skull was infested with parasites); audital bullae smaller and very much narrower; carnassial teeth and upper molar smaller. The skull of the female is very much smaller than that of the male, and has the smoothly rounded brain case of the *cicognani* group, without trace of a sagittal ridge. The squamosals are strongly inflated, resembling those of *cicognani* and the female of *noveboracensis*. It differs from the female *frenatus* in much smaller size, very much smaller audital bullae, more inflated squamosals, smoothly rounded brain case without trace of sagittal crest, and broader interorbital constriction, which is immediately behind postorbital processes instead of one-fifth the distance from the processes to the occipital crest (fig. 15).

Remarks.—On first examining the skins of this weasel sent home by Mr. Nelson, I supposed it to be merely a tropical subspecies of *frenatus;* but on comparing the skulls I am forced to accord it full specific rank. The difference is greatest in the females, and is really very remarkable, as may be seen from the accompanying figures (figs. 15 and 16). The female of *frenatus* (fig. 16) resembles the male of the same species (pl. III, fig. 1), while the female of *tropicalis* (fig. 15) resembles the *cicognani* group—representing another section of the genus. The case is parallel to that of *P. noveboracensis* already described. The female of *tropicalis*, like that of *noveboracensis*, shows arrested development or absence of the specialization that characterizes the male, while the females of *washingtoni* and *frenatus* have advanced further and are more like the male. In the case of the female skulls of *frenatus* and *tropicalis* here figured, it is interesting to know that they were taken within a few miles of one another—*frenatus* on Cofre de Perote, at an altitude of about 12,500 feet; *tropicalis* at Jico on the plain below, at an altitude of 5,000 or 6,000 feet.[1]

The Department collection contains four specimens of this weasel, all collected by Mr. Nelson in Vera Cruz. Three of them, two adult males and one old female, are from Jico; the fourth, an immature female, is from Catemaco, and presents the extreme of differentiation in intensity of color. The hind feet

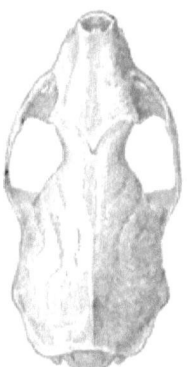 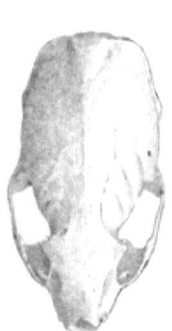

Fig. 15—*P. frenatus* ♂.   Fig. 16.—*P. tropicalis* ♀.

are dark throughout and the color of the upper parts is peculiarly dark and rich, as in *P. affinis*.

Measurements.—Average of two adult males from Jico, Vera Cruz (type locality): Total length, 442; tail vertebræ, 175; hind foot, 50. An old female from same place: Total length, 333; tail vertebræ, 121; hind foot, 37.

PUTORIUS AFFINIS (Gray).

Mustela affinis Gray: Annals & Mag. Nat. Hist., 4th ser., XIV, p. 375, Nov., 1874.

Type locality.—"New Granada" [= Colombia].

General characters.—Size large; tail long; color very dark, almost black anteriorly; facial markings obsolete or nearly so.

Color.—Upper parts nearly pure black on head and neck, fading imperceptibly to rich blackish brown on back, rump, and tail; black tip of tail long, but not strongly contrasted with dark color of rest of tail; under parts narrowly ochraceous orange, narrowest behind angle of mouth, where it is encroached on by the blackish of the cheeks. Face usually unmarked, but a whitish streak sometimes present in front of ear.

Cranial characters.—The only skull of this weasel I have seen is from a skin (No. 13770, U. S. Nat. Mus.) collected by Dr. Van Patten, at San Jose, Costa Rica. It is immature, but differs strikingly from *frenatus* in the greater breadth of the frontal region and the flatness of the audital bullæ. The constriction is little marked, which may be due to

----

[1]The difference in size of the two species is well shown by the flesh measurements of these two specimens. Female *frenatus*, Cofre de Perote: Total length, 418; tail vertebræ, 160; hind foot, 45. Female *tropicalis*, Jico: Total length, 333; tail vertebræ, 121; hind foot, 37.

parasites in the frontal sinuses. The young skull affords the following measurements: Basal length, 50; zygomatic breadth, 29; postpalatal length, 26; palatal length, 24; interorbital breadth, 12; breadth across postorbital processes, 15; breadth of constriction, 14.

General remarks.—There are several specimens from Costa Rica in **the National** Museum collection which apparently belong to this **species.** In these specimens the color of the upper parts is exceedingly dark from the color of the tips of the hairs; but the color immediately underlying the black tips is deep fulvous brown, giving a very rich tone to the pelage. The orange of the under parts is narrow and does not reach the feet; on the hind legs it stops on the thighs, and on the forelegs it stops short of the wrists.

Measurements (from dry skins in U. S. Nat. Mus.).—Total length, about 510; tail vertebrae, about 180; hind foot, about 52.

Table of average cranial measurements of North American Weasels.

Name.	Locality.	Sex.	Basal length.	Basilar length of Hensel.	Zygomatic breadth.	Mastoid breadth.	Breadth across post. orbital processes.	Interorbital breadth.	Foramen magnum to plane of last molars.	Palatal length.	Postpalatal length.	Number of skulls in average.
P. cicognani	Ossipee, N. H.	♂	38.5	37.5	21	18.5	10.5	8.7	25.5	16.5	22	4
	Elk River, Minn.	♂	40.2	39	22	19.5	11	9	26.5	17	22.5	4
	Do.	♀	33.5	32.5	18	16	10	7.8	22	14	19.5	1
	Mount Forest, Ontario.	♀	32.5	31.5	17.5	16	9	7	21.5	14	18	1
P. richardsoni	Great Slave Lake	♂	41.5	40	24	20.5	11.5	9.7	27	18	23	3
P. alascensis	Juneau, Alaska	♂	43	42	24.5	21	11	11	28.5	19	24	1
P. streatori	Skagit Valley, Wash.	♂	35	34	20	18	11	8.5	23	15	20	1
	Do.	♀	30	29.5	16.5	15	10	7.5	20	12	18	1
	Trout Lake, Wash.	♂	33.5	32.5	18	16.5	9.8	8	22	14.5	19.5	3
	Do.	♀	28.5	28	15.5	13.5	8.5	6.5	19	12.5	16.5	1
P. rixosus	Osler, Saskatchewan	♀	26.5	26	14.2	13.5	7.5	5.5	17.5	11	¹15	1
P. arcticus	Point Barrow, Alaska.	♂	44.5	43	29.5	23	14.5	12.5	29	20.5	24	1
	Franklin Bay, Arctic Coast.	♂	43.5	42	27.5	22.5	13	11	28.5	19.5	24	1
	St. Michaels, Alaska	♂	43	42	26.5	22.5	13.5	12	28	19.5	24	1
	Do.	♀	38	37	22.5	19	12	10	24.5	16.5	21	1
P. kadiacensis	Kadiak Island, Alaska	♂	¹42	¹41	24	20.5	12.5	10.5	27	17.5	¹24	1
P. noveboracensis	Adirondacks, N. Y.	♂	47	45.5	27	23.5	14.5	11.3	30	21.5	25.5	5
	Do.	♀	38.5	37.5	20	18.5	11	8.5	25.5	16	22.5	1
P. washingtoni	Trout Lake, Wash.	♂	44.2	43	26	23	12.5	10.2	27.5	21	23	7
	Do.	♀	38.3	37.5	21.5	20	10.5	8.7	24	17.8	20.5	4
P. peninsulae	Tarpon Springs, Fla.	♀	45.5	44	27	24	14	11	29	21	24.5	1
P. longicauda	Carlton House, Saskatchewan.	♂	48	47	30.5	26	15.5	11.5	30	23	25	1
	Do.	♀	43.5	42.5	26	23	12	10.5	26.5	20.5	22.5	1
P. spadix	Elk River, Minn.	♂	48	46.5	29.5	26	14.5	11.5	30	23.5	24	1
	Do.	♀	44	43	26	23.5	13	10.5	28	20.5	23.5	1
P. saturatus	Siskiyou Mountains, Oregon.	♂	45	44	29	25	14	11	28.5	21	24	2
P. arizonensis	Springerville, Ariz.	♂	42	41	26	23	12.5	10.5	26.5	20	22	1
	Boulder County, Colo.	♂	44	43	28.5	23	13	10.5	27	20.5	23.5	1
	Sierra Nevada, Cal.	♂	44.5	43.5	28	23	14.5	10.5	28	21	23	4
	Do.	♀	39.5	38	22.5	20.5	11.3	9	24	18	21	3
P. alleni	Black Hills, S. Dak.	♂	42	40.8	27	22	13.2	11	26.2	20	22	2
	Do.	♀	38.5	37.5	23	....	12	9	24.5	18	20.5	1
P. xanthogenys	Southern California	♂	44	42.5	27.5	23.5	13.5	9.5	27.5	20.5	23.2	2
	Do.	♀	42	41	24	22.5	12	9.5	26	19.5	22	1
P. frenatus	Tlalpam, Mexico.	♂	52.5	51	33.5	27.5	15.5	12	33.5	24.5	27.5	1
	Cofre de Perote, Mexico.	♀	45	43.5	25.5	23	13	10	29	19.5	25	1
P. tropicalis	Jico, Vera Cruz, Mexico.	♂	49	47.5	28	24.5	15	10.5	32	22	27	1
	Do.	♀	37.5	36.5	22.5	19.5	12	9	24.5	16	21.5	1

¹ Estimated.

INDEX.

35

36

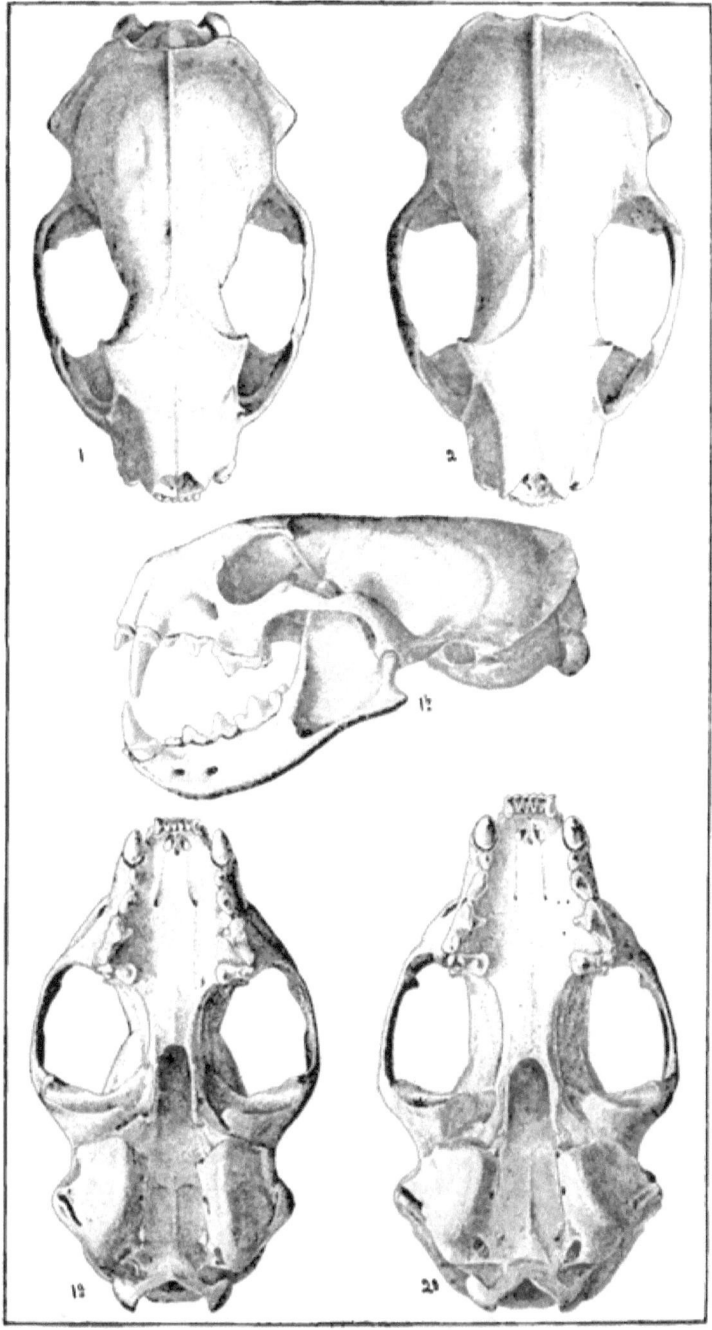

1. *Putorius nigripes* ♂ ad. Trego County, Kansas.
2. *Putorius putorius* ♂ ad. Brunswick, Germany.

PLATE II.

FIG.   1. *Putorius arcticus.*   Point Barrow, Alaska (type).
♂ ad., No, 23010, U. S. Nat. Mus.
2. *Putorius alascensis.*   Juneau, Alaska (type).
♂ ad., No. 74423, U. S. Nat. Mus., Dept. Agric. coll.
3 and 4. *Putorius cicognani.*
3. ♂ ad., Bucksport, Me., No. 4247, Bangs coll.
4. ♀ ad., Mount Forest, Ontario, No. 789, Bangs coll.
5 and 6. *Putorius streatori.*   Mount Vernon, Skagit Valley, Wash.
5. ♂ ad., No. 76646, U. S. Nat. Mus., Dept. Agric. coll. (type).
6. ♀ ad., No. 76623, U. S. Nat. Mus., Dept. Agric. coll.
7. *Putorius rixosus.*   Osler, Saskatchewan.
♀ ad., No. 642, Bangs coll. (type).

38

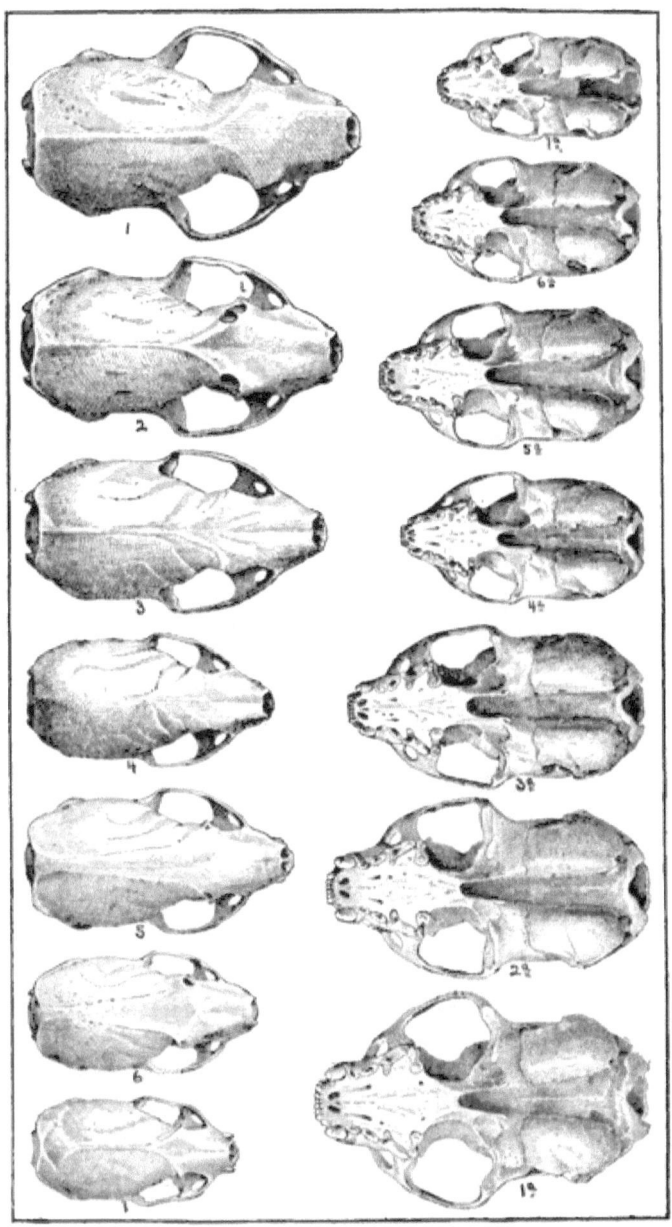

1. *Putorius arcticus.*   3, 4. *P. cicognani.*
2. *P. alascensis.*   5, 6. *P. streatori.*
          7. *P. rixosus.*

PLATE III.

FIGS. 1 and 2. *Putorius frenatus.*

    1. ♂ ad., Tlalpam, Mexico, No. 50826, U. S. Nat. Mus., Dept. Agric. coll.

    2. ♀ ad., Cofre de Perote, Vera Cruz, Mexico, No. 54278, U. S. Nat. Mus., Dept. Agric. coll.

3 and 4. *Putorius longicauda.* Carlton House, Saskatchewan (type locality).

    3. ♂ ad., No. 73183, U. S. Nat. Mus., Dept. Agric. coll.

    4. ♀ ad., No. 75483, U. S. Nat. Mus., Dept. Agric. coll.

5 and 6. *Putorius tropicalis.* Jico, Vera Cruz, Mexico.

    5. ♂ ad., No. 54994, U. S. Nat. Mus., Dept. Agric. coll. (type).

    6. ♀ ad., No. 54993, U. S. Nat. Mus., Dept. Agric. coll.

40

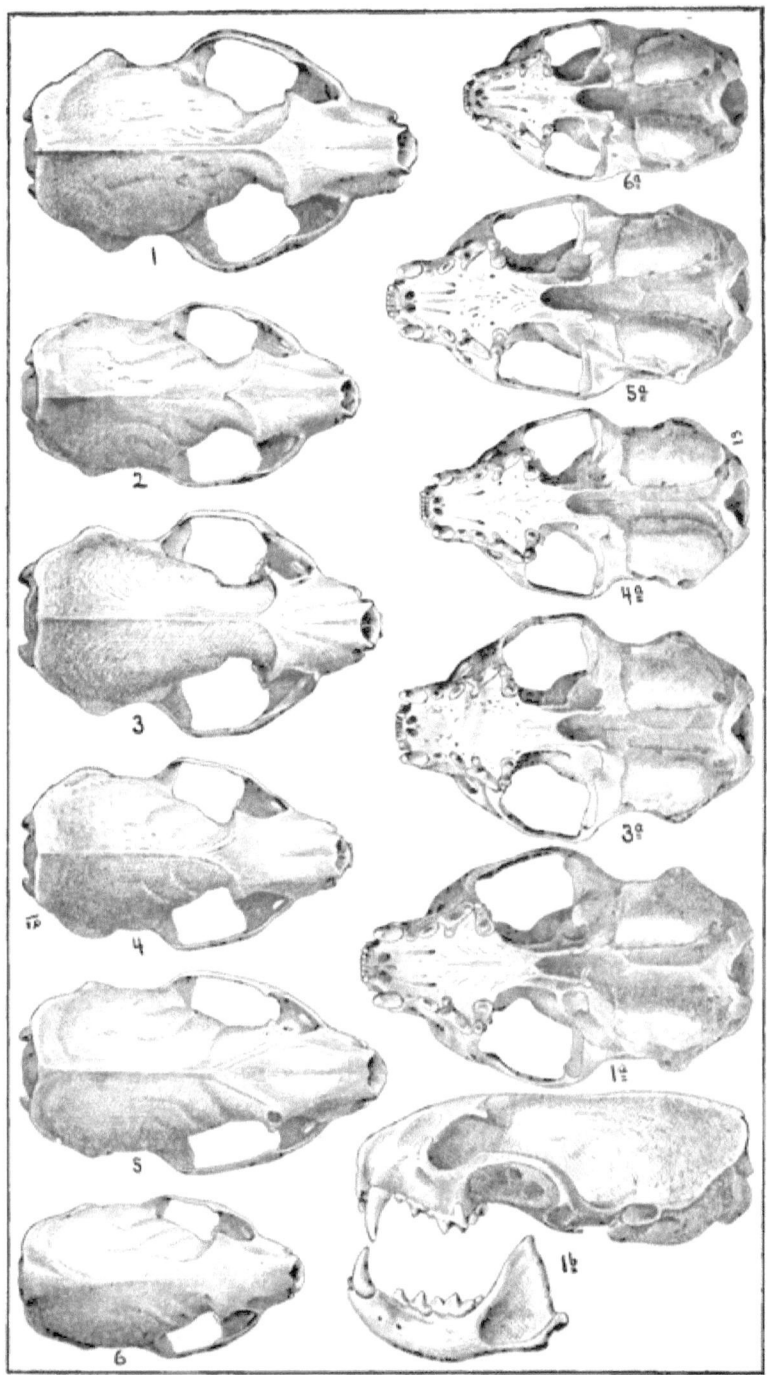

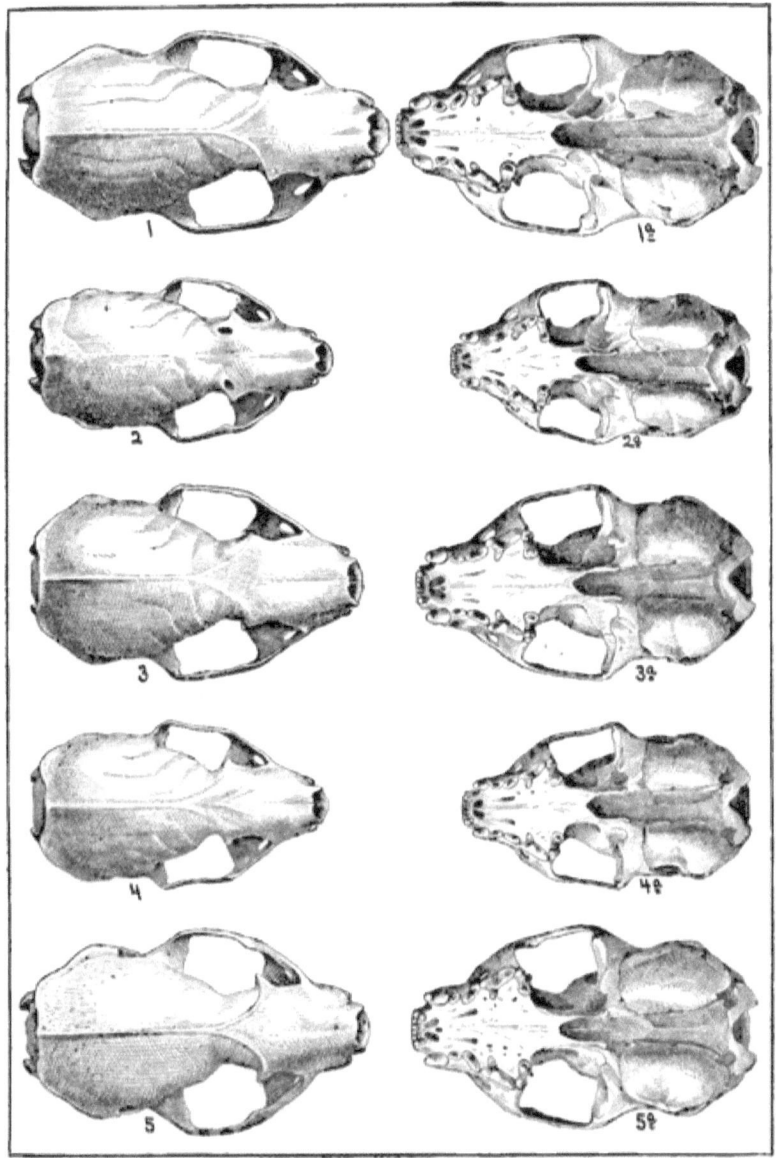

1, 2. *Putorius noveboracensis.*   3, 4. *P. washingtoni.*   5. *P. peninsula.*

PLATE V.

Fig. 1. *Putorius longicauda* (Bonap.).

    1. ♂ ad., Carlton House, Saskatchewan, No. 73183, U. S. Nat. Mus., Dept. Agric. coll.

    1a. ♀ ad., Carlton House, Saskatchewan, No. 75483, U. S. Nat. Mus., Dept. Agric. coll.

2. *Putorius cicognani* (Bonap.).

    2. ♂, Bucksport, Me. No. 4247, Bangs coll.

    2a. ♀, Mount Forest, Ontario No. 789, Bangs coll.

3. *Putorius noveboracensis* De Kay.

    3. ♂ ad., Adirondacks, New York No. 3843, Merriam coll.

    3a. ♀ ad., Adirondacks, New York No. 5598, Merriam coll.

4. *Putorius rixosus* nob.

    ♀ ad. (type), Osler, Saskatchewan, No. 642, Bangs coll.

5. *Putorius peninsulæ* Rhoads.

    ♀ old, Tarpon Springs, Fla. No. 2379, Rhoads coll.

6. **Putorius arcticus** sp. nov.

    6. ♂, St. Michaels, Alaska No. 36243, U. S. Nat. Mus.

    6a. ♀, St. Michaels, Alaska No. 36246, U. S. Nat. Mus.

44

PLATE V.

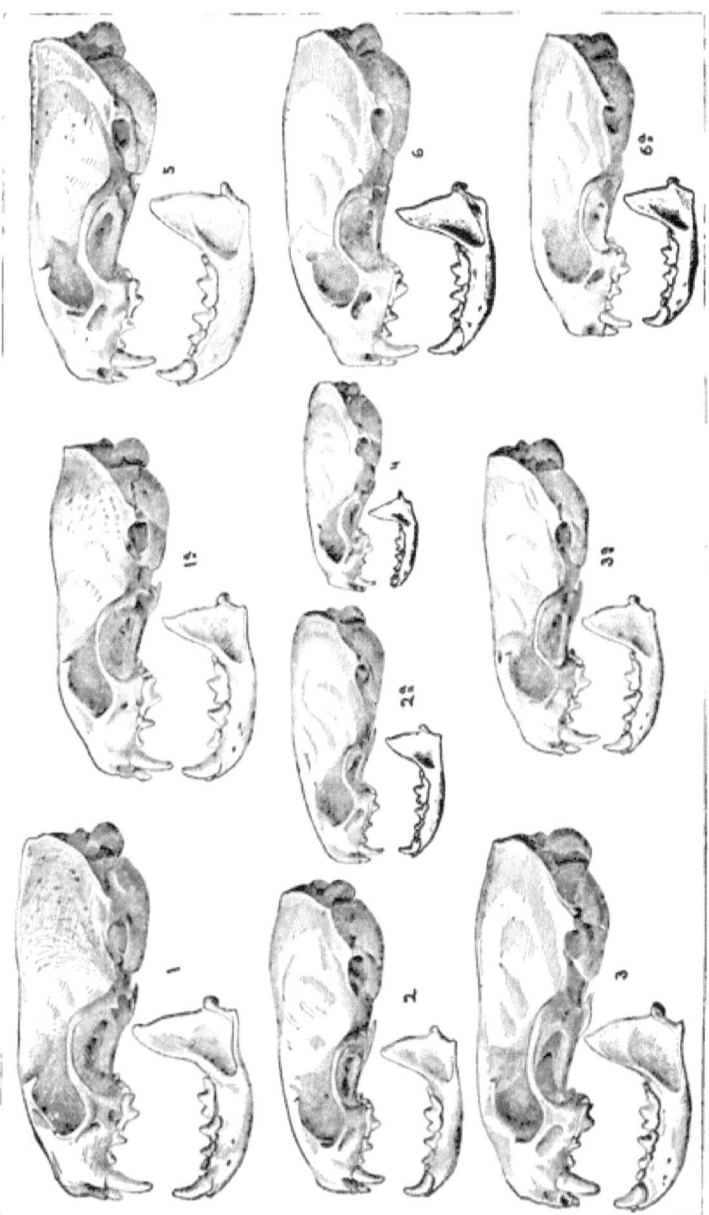

1. *Putorius longicauda.*   2. *P. cicognani*   3. *P. noveboracensis.*   4. *P. rixosus.*   5. *P. peninsulæ.*   6. *P. arcticus.*

www.ingramcontent.com/pod-product-compliance
Lightning Source LLC
Chambersburg PA
CBHW031322280626
47169CB00019B/2617